Mrs. Tung's Qipaos

董堪宇　编

Editor　Kenneth Tung

母親的旗袍

上海書店出版社
SHANGHAI BOOKSTORE PUBLISHING HOUSE

谨以此书纪念我敬爱的母亲

This book is dedicated to the memory of my beloved mother

目 录

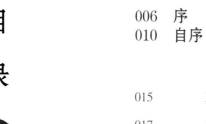

CONTENTS

母亲的旗袍

Mrs. Tung's Qipaos

序

这本相册汇集的是一位百岁老人——董胡修凤女士的旗袍图集。这些旗袍是她老人家百年之后，她的儿子董堪宇先生捐献给我们旗袍馆的。

在上海滩以旗袍风采著名的女性中，董胡修凤女士或许不算太知名，但是她绝对是其中非常特殊的一位。她是老一代民族实业家、当年著名的上海统益纺织股份有限公司董事兼总经理、上海市棉纺织工业公司副经理、上海市政协数届常务委员、上海民建数届常务委员董春芳先生的夫人。她与丈夫相濡以沫、风雨同舟，共同走过了 54 个风云变幻的难忘年头。这样的身份，很容易使人联想起，上海滩十里洋场的珠光宝气和大家闺秀的仪态万方。可是几年前，当我们去上海华侨新邨拜访老人家时，她那年已经 97 岁高龄，正在她那宽敞典雅的客厅里，聚精会神地观看电视新闻报导。

当她以微笑向我们打招呼时，我们觉得眼前是位七十来岁的老太太。她雍容文雅，身手敏捷，衣衫楚楚，毫发不乱，尤其那双明亮的眼睛，充满智慧又善解人意。老人家话语不多，慢条斯理，每句话都"拿捏"得很有分寸。这无疑"迫使"我们不得不收敛一下平时的"野性"，尽可能变得文质彬彬地……我曾先后三次访问老太太，其中一次是在她美国加州奥克兰的湖边公寓。愚钝如我，渐渐明白了，老太太是一位大上海百年风云中，历尽沧桑而能从容淡定、任凭窗外风起云落，始终能顺其自然的超级智者。她未曾在社会上任职，但却实际充当了十分重要的角色——老太爷的贤内助……自然，这不是本篇短文能够详叙的。

面对老太太这定做于 20 世纪 80 年代初期的 30 件旗袍，参见她 40 年来身穿这些旗袍的生活照片，似乎感觉这些旗袍都有温度和气质了。看来，她比较喜欢薄如蝉翼的真丝面料和清纯淡雅的花卉图案，玫瑰花、月季花、丁香花、康乃馨、菊花……除了两件单色的旗袍之外，全是带有花卉和图案的印花面料。那些花卉，绝大多数是在淡绿、淡蓝、淡紫、淡粉，或淡红的底色上，似工笔画一般，枝枝叶叶，细细描出，除了两件丝绒面料的深色大花，其余均以素淡胜出。当时上海一批旧时代的老裁缝也还在世，那些细腻缜密的针脚，柔韧适度的绲边，高低得体的开衩，无不记录了当年旗袍制作工艺的严谨和规范，也显示了老人家对旗袍一丝不苟的制作要求。

声如其人，衣也如其人。服饰显示的是一个人的修养和品位。我们从老人家的旗袍中，一再读出的是清新、淡雅、平和、纯真。这是她的旗袍传递出来的弦外之音。我们有幸得以珍藏这批美丽的旗袍，也有幸从中获取旗袍之外的精神营养，何其美哉！我们理应传承好这种美丽而高雅的海派精神，不辜负前辈们的期望。

再次感谢董堪宇先生的信任。

是为序。

上海老旗袍珍品馆副馆长

宋路霞

2021 年 4 月 5 日

FOREWORD

This volume presents a pictorial survey of the collection of qipaos belonging to Mrs. Shu Feng Hu Tung, who lived to 100. After her death, her son, Mr. Kenneth Tung, donated these fine dresses to us at the Shanghai Pavilion of Treasured Qipaos.

Mrs. Shu Feng Hu Tung may not have stood out from the qipao-clad ladies moving in the elite circles of old Shanghai, but she was a highly unusual character nonetheless. She was married to Mr. Chung Fong Tung, an industrialist with a notable array of titles. He was a board member and general manager of the Shanghai Tung Yih Cotton Mill, Ltd., and subsequently the deputy manager of the Shanghai Cotton Textile Industrial Co. He also served on the standing committee of both the CPPCC Shanghai Committee and the China National Democratic Construction Association Shanghai Committee for several consecutive terms. Throughout all 54 eventful years of their life together, the couple stood steadfastly by each other through thick and thin. One is led to believe that she was one of their formidable bejeweled grand society ladies of old Shanghai. But when we called on her a few years ago, the 97-year-old was watching news on the television in her spacious, charming living room, a far cry from a "living legend" that we half expected.

As the beaming lady came up to greet us, we thought she was still in her seventies — graceful, elegant, agile, well dressed, and impeccably groomed. She had striking bright eyes brimming with wisdom and understanding. She wasn't given to words, but everything she had to say was carefully weighed. This forced us to rein in our usual liveliness and act as "proper" as we could manage. I paid two more visits — once in her Lakeside Drive apartment in Oakland, California. With time, I came to appreciate her calm and collected sage-like presence, belying a life of strife and survival amidst the sweeping social changes that had rocked Shanghai over the past century. She never held any official position, but she played a critical role as a supportive wife of her husband. Of course, I don't have space to go into all this here.

Looking at Mrs. Hu Tung's 30 qipaos tailored in the early 1980s, alongside the photos of her in them dating from her last four decades of life, I was struck by the ineffable aura emanating from these wonderful dresses. It seems to me that she preferred paper-thin silk fabrics and simple, tasteful floral patterns, such as roses, lilacs, carnations, and chrysanthemums. With the exception of two in solid color, all the dresses were crafted from floral patterned printed fabrics. Mostly, the flowers are set against backgrounds of a pastel shade of green, blue, purple, pink, or red, exuding a refinement reminiscent of fine-brush Chinese paintings. The branches and leaves are carefully limned. Except for two velvet dresses with large, dark flowers, they are characterized by their understatement and simplicity. Created at a time when a roster of seasoned tailors were still around, the dresses still impress with their delicate and meticulous stitches, pliable edging, and well-judged slits, channeling the exquisite qipao craftsmanship of yore and Mrs. Hu Tung's discerning taste.

One's sartorial choice reflects her character, manners, and taste. Mrs. Hu Tung's qipaos speak to us across time with all their rich "overtones" that bring to life her unique qualities of simplicity, elegance, calm, and purity. We are fortunate to be able to acquire these beautiful dresses and to live Mrs. Hu Tung's life vicariously through them. It is a beautiful legacy to be treasured.

In closing, I'd like to thank Mr. Kenneth Tung again for his trust.

I hope these words serve well as the Foreword.

Deputy Director
Shanghai Pavilion of Treasured Qipaos
Song Luxia
April 5, 2021

自序

我的养母董胡修凤，生于 1919 年，原籍浙江舟山，是家父董春芳先生（上海老一代民族棉纱工商业者）的夫人，享年 100 岁。1979 年家父移居美国，1980 年养母赴美，他们时常往返中美两地。她老人家一生的经历证明了她不同凡响的智慧、情感、品行和贤惠以及对家父的崇敬、爱戴和关怀。

养母历来淡妆素面，从不浓妆艳抹，服饰穿戴清秀高雅，尤其爱穿旗袍，旗袍服装为她的生活增添了无限情趣和艺术雅韵。

养母穿旗袍经常穿旗袍套装或以披肩作衬配，款式不受年代流行的限定。她爱用各种优质轻薄的真丝印花或提花面料，很少用锦缎或毛料，一般选用色彩清淡优雅的散花或图案，不求珠光宝气，落落大方，偶尔选择鲜艳夸张和对比度大的花色。她的旗袍门襟都用暗扣（揿钮）而不用盘扣。她认为旗袍做工一定要量身定做才能服帖，针脚要细巧，衣领高低和领口大小、袖长和袖口的大小、盘扣和绲边的设计、衩口长短和下摆尺寸的合度都是必不可少的条件，而旗袍的长短和色彩更要视不同的料子以及所穿的时间和场合而定。

她的旗袍长度都在膝盖以下而露出脚踝，她认为必须相配与旗袍色彩谐调的长筒丝袜和高跟鞋，才能真正体现中国妇女穿旗袍时的身姿风韵。由于无法找回她的老旗袍，本册所编入的旗袍都是她在 1980 年代前后，由上海的数位私人裁缝为她量身定做的。

养母的一切都对我意义非凡，最重要的是她对我一生的谆谆教导，成功地把我培养成一位有志气有追求的人。她老人家善于言教身教，教我待人接物、识明审智、识才尊贤、为人正直、助人为乐、发奋读书、知足常乐。为此，我庆幸虽然自幼娇生惯养，竟没有沦为纨绔子弟而碌碌无为，从国内一个 1967 年毕业的高中生，到美国后四年中完成学业，成为我们家庭中第一个毕业于美国大学并列入优秀学院学生名单的学士和硕士，曾在美国数家首列大公司任职，直到退休。

丁酉年（2017 年）上海龙华寺高僧海晏师父赋养母"福满百年"四字。2019 年正是她福满百年，去冰岛邮轮旅游庆贺她百岁生日之后，不久安详驾鹤西游。为了表达我对她养育之恩的深刻感激和无限爱戴及思念，我选出了养母的 30 件旗袍，捐赠予"上海老旗袍珍品馆"永久收藏，以支持传承中国妇女高雅无俦的旗袍文化，希望中国的旗袍文化万代留世。

兹编辑此影集与世人共享。

董堪宇

2021 年 3 月

PREFACE

My adoptive mother, Shu Feng Hu Tung, was born in 1919. A native of Zhoushan, Zhejiang province, she married Chung Fong Tung, a cotton-mill industrialist in pre-1949 Shanghai. She lived to the venerated age of 100. After my father moved to the United States in 1979, my adoptive mother followed in 1980. Since then, they had often traveled back and forth between the two countries. Her life is a witness to her extraordinary wisdom, temperament, character and virtues, and her reverence, love and care for my father.

My adoptive mother always kept her look low-key and subtle. She was always elegantly clad, with a special preference for the qipao, a dress that lent much interest and flair to her everyday life.

She often wore her qipao with a matching top or a shawl, cultivating a look that didn't necessarily follow the trends of the day. She loved all sorts of fine light printed silk or jacquard fabrics — enlivened by floral motifs or other patterns in elegant colors — rather than brocade or wool. Never aiming for a majestic or regal presence, she always looked poised and tasteful. Occasionally, though, she would take on bright designs, or those rich in contrasts. The front of her qipaos is fastened with hidden snaps, rather than pankous. For her, the well-fitting qipao should be tailored to one's shape; as important are the stitching, the height and size of the collar, and the length of the sleeves and the fit of the cuffs. The design of the pankous, edging, the length of the side slits and the width of the hem should all be carefully calibrated, while the length and color of the qipao may vary according to the fabrics used and the occasions it is created for.

Her hemline of qipaos reaches between her knees and ankles. She believes that, to showcase the Chinese ladies at their graceful best, these exquisite dresses must be worn with silk stockings and high heels. As she had lost trace of most of her qipaos from the old days, she had them recreated by several private tailors in Shanghai around the 1980s. These are the dresses featured in this book.

Everything my adoptive mother meant a lot to me, but the most important thing she ever did for me was her earnest guidance. This proved critical to my development and instilled ambition and aspiration in me. She instructed and influenced me with her words and deeds, showing me how to get along with others and make decisions, how to be upright, helpful, diligent, and be content with moderate conditions. The education prevented me from becoming pampered and falling behind: I completed my tertiary education within four years in the US despite my background as a school leaver in 1967. I am the first in our family to graduate from a US university, receiving both a bachelor's and a master's degree, and making the Dean's List several times while an undergraduate. I later worked my way into leading companies until my retirement.

In 2017, Master Haiyan, a senior Buddhist monk from Shanghai's Longhua Temple, dedicated a calligraphic inscription to my adoptive mother that read "Blessings for a full century." It was a fitting souvenir to celebrate her centenary birthday in 2019, but she died a peaceful death shortly after her Icelandic cruise tour that year. In order to express my admiration and deep gratitude for her love, I selected 30 qipaos of hers and donated them to the permanent collection of the Shanghai Pavilion of Treasured Qipaos. I hope that the quintessential Chinese qipao culture can live on and never perish.

I hope that you will enjoy reading this volume.

Kenneth Tung
March 2021

1980 年美国俄亥俄州托莱多市的寓所
Apartment in Toledo, Ohio, 1980

真丝黑底印咖啡色大小花卉短袖旗袍

Silk short-sleeve qipao with brown floral motifs on black

1981 年和父亲在美国朋友家
At a friend's house, 1981

真丝粉底印大小蓝花无袖旗袍

Silk sleeveless qipao with large and small blue floral patterns on pink

Mrs. Tung's Qipaos

1982 年美国加州奥克兰市的寓所
Apartment in Oakland,
California, 1982

2007 年澳洲邮轮之旅
Australia cruise, 2007

2007 年澳洲邮轮之旅和我合影
Australia cruise, 2007

真丝黑底印花白色大牡丹无袖旗袍

Silk sleeveless qipao with white large peony patterns on black

1982 年美国加州奥克兰市的寓所
Apartment in Oakland,
California, 1982

2005 年伦敦邮轮之旅
London cruise, 2005

2009 年地中海邮轮之旅
Mediterranean cruise, 2009

真丝褐色底色树叶与玫瑰印花短袖旗袍

Silk short-sleeve qipao with leaves and rose patterns on brown

1982 年同父亲在美国亚利桑那州凤凰城看望我

Visited me in Phoenix, Arizona in 1982

真丝藕色甲骨文碎花印花短袖旗袍

Silk pinkish grey short-sleeve qipao with floral and oracle-bone patterns

1986 年同父亲在美国纽约市参加我的婚礼

At my wedding in New York City, 1986

黄色织锦缎菊花花纹旗袍套装

Yellow brocaded qipao suit with chrysanthemum patterns

1988 年美国加州奥克兰市的寓所，父亲所摄
Apartment in Oakland, California, picture taken by Father, 1988

2006 年北欧邮轮之旅
Scandinavia cruise, 2006

真丝黑底牡丹大花印花短袖旗袍

Silk short-sleeve qipao with peony patterns on black

1996 年旧金山父亲 90 岁的生日宴会
At Father's 90th birthday party in San Francisco, 1996

丝绒红蓝印花大花短袖旗袍套装

Velvet short-sleeve qipao with large red-and-blue floral patterns with a black top

2005 年伦敦邮轮之旅
London cruise, 2005

丝绒红蓝印花大花无袖旗袍

Velvet sleeveless qipao with large red-and-blue floral patterns

2006 年北欧邮轮之旅
Scandinavia cruise, 2006

2007 年澳洲邮轮之旅
Australia cruise, 2007

2007 年美国加州爱莫利维尔市的生日宴会
Birthday party in Emeryville, California, 2007

2010 年东加勒比海邮轮之旅
Eastern Caribbean cruise, 2010

真丝绿底黑色钟鼎文花纹印花旗袍套装

Silk qipao suit with black bronze inscription patterns on green

2007 年西加勒比海邮轮之旅
Western Caribbean cruise, 2007

真丝咖啡色底印黑色小花短袖旗袍

Silk brown short-sleeve qipao with small black dots on brown

母亲的旗袍　Mrs. Tung's Qipaos

2007 年西加勒比海邮轮之旅
Western Caribbean cruise, 2007

2010 年东加勒比海邮轮之旅
Eastern Caribbean cruise, 2010

真丝灰底黑色小花印花短袖旗袍

Silk short-sleeve qipao with black small floral patterns on grey

2007 年澳洲邮轮之旅
Australia cruise, 2007

真丝白底绿花印花无袖旗袍

Silk sleeveless qipao with green floral patterns on white

2007 年西加勒比海邮轮之旅
Western Caribbean cruise, 2007

2008 年巴拿马运河邮轮之旅
Panama Canal cruise, 2008

真丝绿色树叶花纹印花短袖旗袍

Silk short-sleeve qipao with leaf patterns on green

2007 年澳洲邮轮之旅
Australia cruise, 2007

2007 年澳洲邮轮之旅　　　　2009 年地中海邮轮之旅　　　　2011 年东南亚邮轮之旅
Australia cruise, 2007　　　Mediterranean cruise, 2009　　Southeast Asia cruise, 2011

黄色织锦缎旗袍配蕾丝外套（旗袍套装）

Yellow brocaded qipao with a lace top

2007 年澳洲邮轮之旅
Australia cruise, 2007

2007 年澳洲邮轮之旅和我合影
Australia cruise, 2007

真丝浅蓝底印大小蓝花短袖旗袍

Silk short-sleeve qipao with large and small blue floral patterns

2009 年地中海邮轮之旅和我合影

Mediterranean cruise, 2009

深蓝黑底丁香花纹印花短袖旗袍

Silk short-sleeve qipao with lilac patterns on dark blue

2009 年地中海邮轮之旅
Mediterranean cruise, 2009

真丝红黑杂花印花与提花短袖旗袍

Silk short-sleeve qipao with red and black patterns

母亲的旗袍

Mrs. Tung's Qipaos

— 049

2013 年阿拉斯加邮轮之旅
Alaska cruise, 2013

真丝提花印花五彩水纹短袖旗袍

Silk short-sleeve qipao with multi-color wave patterns on jacquard print

2014 年在美国华盛顿参加我女儿婚礼
At my daughter's wedding in
Washington D.C., 2014

2017 香港邮轮之旅
Hong Kong cruise, 2017

2017 年上海的生日宴会
Birthday party in Shanghai, 2017

2017 南美邮轮之旅
South America cruise, 2017

真丝仿虎皮花纹印花旗袍套装

Silk qipao suit with tiger skin patterns

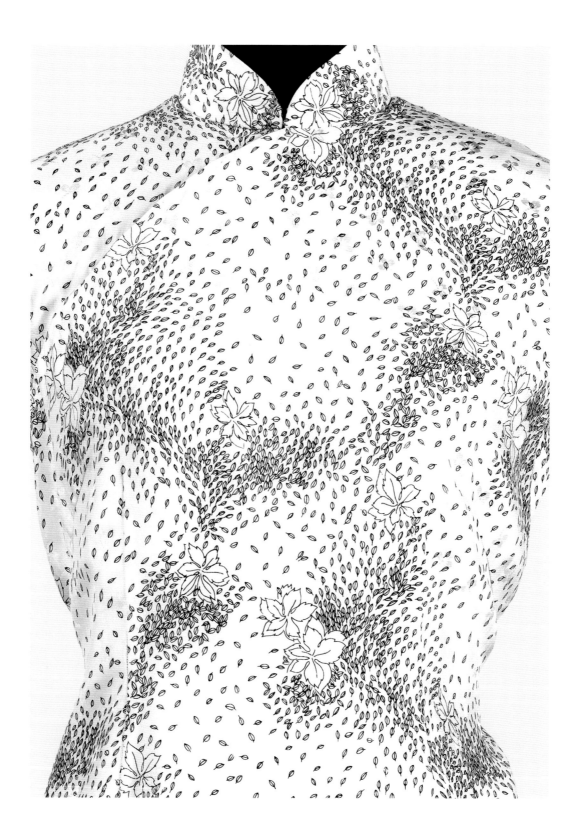

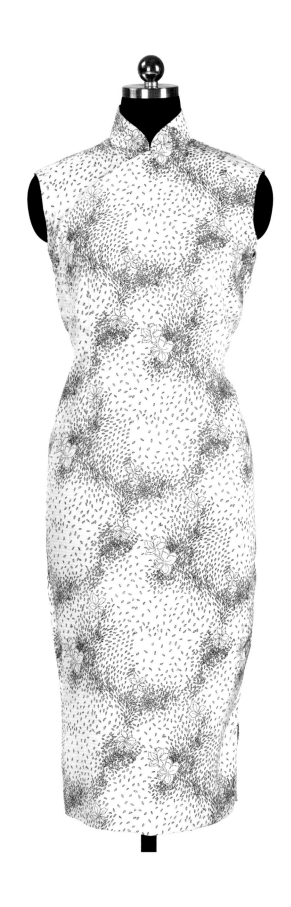

真丝白底碎花印花无袖旗袍

Silk sleeveless qipao with floral patterns on white

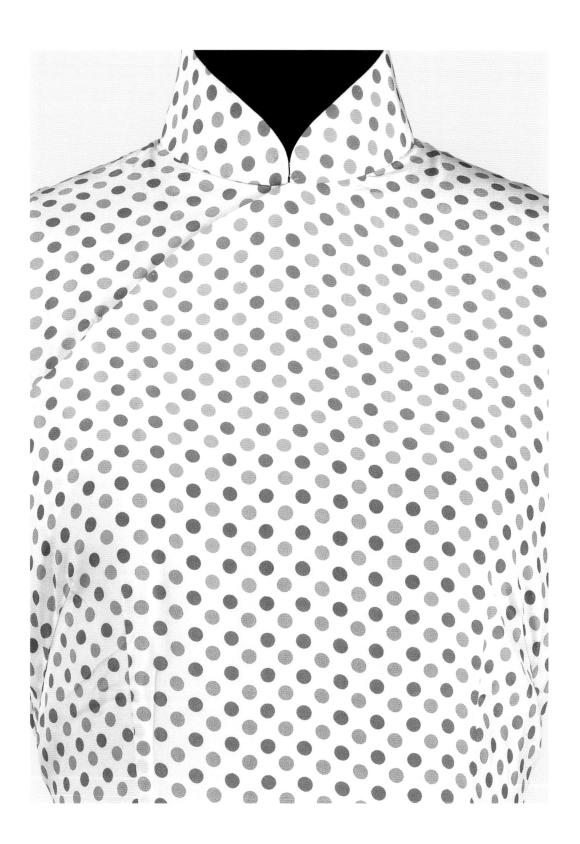

真丝白底灰黄色圆点印花无袖旗袍

Silk sleeveless qipao with grey and yellow polka dots on white

真丝灰底红黑杂花印花与提花短袖旗袍

Silk short-sleeve qipao with mixed red and black patterns on grey

斜纹毛料咖啡色暗格中袖旗袍

Twill wool medium-sleeve qipao with brown dark checkers

黑底真丝碎花印花短袖旗袍

Silk short-sleeve qipao with floral patterns on black

真丝藕色大花印花短袖旗袍

Silk pinkish grey short-sleeve qipao with large floral patterns

真丝蓝底杂花方格印花短袖旗袍

Silk checkered short-sleeve qipao with mixed floral patterns

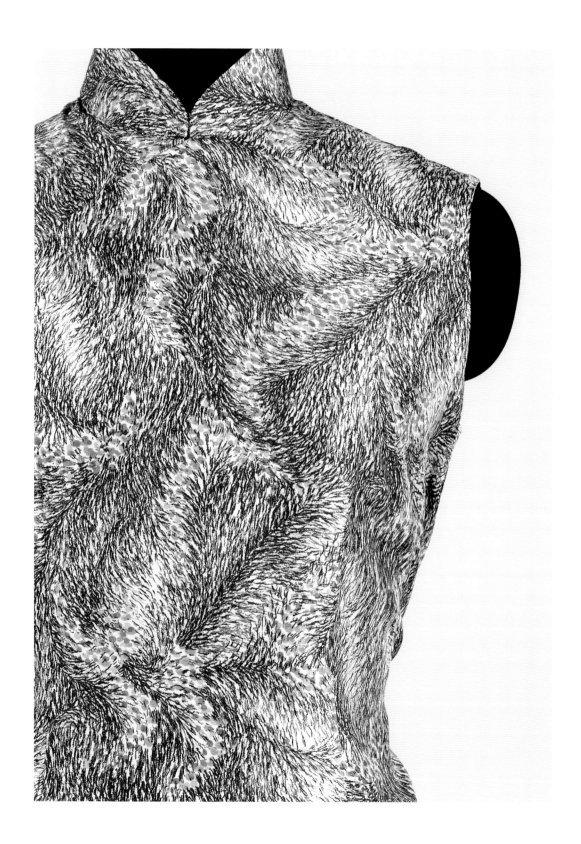

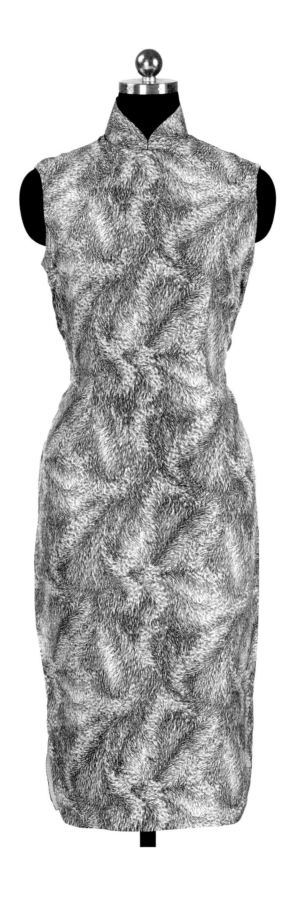

真丝羽毛花纹印花无袖旗袍

Silk sleeveless qipao in feather print

全黑毛料长袖旗袍

All-black long-sleeve wool qipao

橙色毛料中袖旗袍

Orange wool medium-sleeve qipao

图书在版编目（CIP）数据

母亲的旗袍 : 汉、英 / （美）董堪宇编 . -- 上海：
上海书店出版社，2021.10
ISBN 978-7-5458-2099-7

Ⅰ . ①母… Ⅱ . ①董… Ⅲ . ①旗袍—中国—图集
Ⅳ . ① TS941.717.8

中国版本图书馆 CIP 数据核字 (2021) 第 195870 号

责任编辑　章玲云
顾　　问　宋路霞

母亲的旗袍
董堪宇 编

出　　版　上海书店出版社
　　　　　　（201101　上海市闵行区号景路159弄C座）
发　　行　上海人民出版社发行中心
印　　刷　上海雅昌艺术印刷有限公司
开　　本　889×1194　1/16
印　　张　4.5
版　　次　2021年10月第1版
印　　次　2021年10月第1次印刷
ISBN 978-7-5458-2099-7/TS.22
定　　价　180.00元